Energy Sector Standard of the People's Republic of China

NB/T 10145-2019

Regulation for preparation of final account report for hydropower projects

水电工程竣工决算报告编制规定

(English Translation)

China Water & Power Press

Beijing 2024

All rights reserved. No part of this publication may be reproduced, stored in a retrieval system, or transmitted in any form or by any means—electronic, mechanical, photocopying, recording or otherwise, without prior written permission of the publisher.

图书在版编目（CIP）数据

水电工程竣工决算报告编制规定 : NB/T 10145-2019 = Regulation for Preparation of Final Account Report for Hydropower Projects (NB/T 10145-2019) : 英文 / 国家能源局发布. -- 北京 : 中国水利水电出版社, 2024. 5. -- ISBN 978-7-5226-2679-6

Ⅰ. TV512

中国国家版本馆CIP数据核字第20248VD863号

Energy Sector Standard of the People's Republic of China

中华人民共和国能源行业标准

Regulation for preparation of final account report for hydropower projects

水电工程竣工决算报告编制规定

NB/T 10145-2019

(English Translation)

Issued by National Energy Administration of the People's Republic of China
国家能源局　发布
Translation organized by China Renewable Energy Engineering Institute
水电水利规划设计总院　组织翻译
Published by China Water & Power Press
中国水利水电出版社　出版发行
　　　Tel: (+ 86 10) 68545888　68545874
　　　sales@mwr.gov.cn
　　　Account name: China Water & Power Press
　　　Address: No.1, Yuyuantan Nanlu, Haidian District, Beijing 100038, China
　　　http://www.waterpub.com.cn
中国水利水电出版社微机排版中心　排版
北京中献拓方科技发展有限公司　印刷
210mm×297mm　16开本　5.75印张　182千字
2024年5月第1版　2024年5月第1次印刷

Price（定价）：￥900.00

About English Translation

This English version is one of China's energy sector standard series in English. Its translation was organized by China Renewable Energy Engineering Institute authorized by National Energy Administration of the People's Republic of China in compliance with relevant procedures and stipulations. This English version was issued by National Energy Administration of the People's Republic of China in Announcement [2023] No.4 dated May 26, 2023.

This version was translated from the Chinese Standard NB/T 10145-2019, *Regulation for preparation of final account report for hydropower projects*, published by China Water & Power Press. The copyright is reserved by National Energy Administration of the People's Republic of China. In the event of any discrepancy in the implementation, the Chinese version shall prevail.

Many thanks go to the staff from the relevant standard development organizations and those who have provided generous assistance in the translation and review process.

For further improvement of the English version, any comments and suggestions are welcome and should be addressed to:

China Renewable Energy Engineering Institute
No. 2 Beixiaojie, Liupukang, Xicheng District, Beijing 100120, China
Website: www.creei.cn

Translating organizations:

POWERCHINA Huadong Engineering Corporation Limited

China Renewable Energy Engineering Institute (China Renewable Energy Engineering Cost Management Center)

Translating staff:

ZHAO Mingzhi	LI Dongwei	Chen Xinkun	LU Yeqi
CHU Kaimin	LIU Qiushi	ZHANG Qinyu	ZHANG Yumin
DAI Honghui	GONG Bowen	LYU Minglu	LIN Lewei
HUANG Deqiang	HU Bin	MOU Xuhui	GAO Shusen
XU Runzhe			

Review panel members:

QIE Chunsheng	Senior English Translator
GUO Jie	POWERCHINA Beijing Engineering Corporation Limited
ZHANG Qing	POWERCHINA Zhongnan Engineering Corporation Limited
LI Kejia	POWERCHINA Northwest Engineering Corporation Limited
QIAO Peng	POWERCHINA Northwest Engineering Corporation Limited
JIA Haibo	POWERCHINA Kunming Engineering Corporation Limited
Wang Yonggang	POWERCHINA Northwest Engineering Corporation Limited
LI Jiewen	China Renewable Energy Engineering Institute

LIU Rongli	China Renewable Energy Engineering Institute
LI Shisheng	China Renewable Energy Engineering Institute

National Energy Administration of the People's Republic of China

翻译出版说明

本译本为国家能源局委托水电水利规划设计总院按照有关程序和规定，统一组织翻译的能源行业标准英文版系列译本之一。2023年5月26日，国家能源局以2023年第4号公告予以公布。

本译本是根据中国水利水电出版社出版的《水电工程竣工决算报告编制规定》NB/T 10145—2019翻译的，著作权归国家能源局所有。在使用过程中，如出现异议，以中文版为准。

本译本在翻译和审核过程中，本标准编制单位及编制组有关成员给予了积极协助。

为不断提高本译本的质量，欢迎使用者提出意见和建议，并反馈给水电水利规划设计总院。

地址：北京市西城区六铺炕北小街2号
邮编：100120
网址：www.creei.cn

本译本翻译单位：中国电建集团华东勘测设计研究院有限公司
　　　　　　　　水电水利规划设计总院（可再生能源定额站）

本译本翻译人员：赵明志　李东伟　陈昕堃　陆业奇
　　　　　　　　褚凯敏　刘秋实　张钦妤　张羽敏
　　　　　　　　代洪慧　龚博文　吕明璐　林乐伟
　　　　　　　　黄德强　胡　斌　牟旭辉　邰树森
　　　　　　　　徐润哲

本译本审核人员：

郄春生　英语高级翻译

郭　洁　中国电建集团北京勘测设计研究院有限公司

张　庆　中国电建集团中南勘测设计研究院有限公司

李可佳　中国电建集团西北勘测设计研究院有限公司

乔　鹏　中国电建集团西北勘测设计研究院有限公司

贾海波　中国电建集团昆明勘测设计研究院有限公司

王永刚　中国电建集团西北勘测设计研究院有限公司

李杰文　水电水利规划设计总院

刘荣丽　水电水利规划设计总院

李仕胜　水电水利规划设计总院

国家能源局

Contents

Foreword	VII
Introduction	IX
1 Scope	1
2 Normative references	1
3 Terms and definitions	1
4 General provisions	1
5 Composition of final account report	2
5.1 Contents of final account report	2
5.2 Cover of final account report	2
5.3 Photographs of the completed project	2
5.4 Final account introduction	2
5.5 Final account statements	2
5.6 Important documents of final account report	3
6 Final account report preparation	3
6.1 Preparation basis of final account report	3
6.2 Conditions for preparation of final account report	4
6.3 Preparation procedure of final account report	4
6.4 Plan development of final account report	4
6.5 Data collection of final account report	4
6.6 Determination of final account base date	5
6.7 Liquidation of final account	5
6.8 Preparation of final account statements	6
6.9 Preparation of final account introduction	14
Annex A (Normative) **Format of the cover of final account report for hydropower projects**	16
Annex B (Normative) **Format of final account statement sheets for hydropower projects**	17
Annex C (Informative) **Reference table for assets composition and prepaid assets classification for hydropower projects**	71
Annex D (Informative) **Reference table for apportionment of prepaid expenses of hydropower projects**	76
Figure A.1 Format of the cover of final account report	16
Table B.1 Fact sheet of the completed project	17
Table B.2 Final account summary sheet	19
Table B.3 Construction auxiliary works final account sheet	21
Table B.4 Construction works final account sheet	23
Table B.5 Environmental protection and soil and water conservation special works final account sheet	26
Table B.6 Electro-mechanical equipment and installation works final account sheet	28
Table B.7 Hydraulic steel structures and installation works final account sheet	35
Table B.8 Resettlement compensation final account sheet	39
Table B.9 Independent cost final account sheet	40
Table B.10 Transfer of usable assets summary sheet	43
Table B.11 Breakdown of transfer of usable fixed assets (structures and buildings)	44

Table B.12	Breakdown of transfer of usable fixed assets (equipment to be installed)	48
Table B.13	Breakdown of transfer of usable fixed assets (equipment not to be installed)	59
Table B.14	Breakdown of transfer of usable assets (current assets)	60
Table B.15	Breakdown of transfer of usable assets (intangible assets and long-term prepaid expenses)	61
Table B.16	Final financial account sheet	62
Table B.17	Breakdown of yearly funds availability	63
Table B.18	Breakdown of wind-up works and reserve fund	64
Table B.19	Breakdown of prepaid expenses	65
Table B.20	Prepaid expenses apportionment calculation sheet	69
Table B.21	Breakdown of infrastructure expenditures to be written-off	70
Table B.22	Breakdown of expenditures transferred-out	70
Table C.1	Reference table for assets composition and prepaid assets classification	71
Table D.1	Reference table for apportionment of prepaid expenses	76

Foreword

This standard is drafted in accordance with the rules given in the GB/T 1.1-2009 *Directives for standardization—Part 1: Structure and drafting of standards*.

National Energy Administration of the People's Republic of China is in charge of the administration of this standard. China Renewable Energy Engineering Institute has proposed this standard and is responsible for its routine management. Energy Sector Standardization Technical Committee on Hydropower Engineering Economics is responsible for the explanation of specific technical contents. Comments and suggestions in the implementation of this standard should be addressed to:

China Renewable Energy Engineering Institute
No. 2 Beixiaojie, Liupukang, Xicheng District, Beijing 100120, China

Drafting organizations:

China Renewable Energy Engineering Institute (China Renewable Energy Engineering Cost Management Center)
POWERCHINA Huadong Engineering Corporation limited
State Grid Xinyuan Holding Co., Ltd. Technology Center
Fujian Xianyou Pumped Storage Company Ltd.

Chief drafting staff:

DAI Honghui	LI Jiewen	GONG Kui	LUO Shuifang
QIAO Tianxia	ZHAI Haiyan	WANG Yuan	ZHANG Huabin
JIANG Xianyu	ZHAO Mingzhi	CHEN Shanglin	ZHANG Ying
ZHANG Yunfei	XU Xumin	LI Dongwei	WANG Jianhang
WU Xiaoyan			

Introduction

The final account report for a hydropower project is the summary statement prepared by the project legal person in accordance with relevant national policies and sector regulations after the project completion to comprehensively reflect the construction costs, construction results and financial status. It is a written statement confirming the value of the newly increased fixed assets and the usable fixed assets transferred. It serves as an analysis report to check the execution of the budget estimate at design stage and to reflect the actual construction cost and investment returns, and is an important part of the final account acceptance and project completion acceptance documents of the hydropower projects.

The *Administrative Measures for the Acceptance of Hydropower Projects* (GNXN [2011] No. 263) states that, "The project completion acceptance shall be conducted on the basis of the special acceptance of project complex, resettlement, environmental protection, soil and water conservation, fire protection, occupational health and safety, project final account and project archives". In order to facilitate the management of final account acceptance of hydropower projects, standardize the preparation of the hydropower projects final account report, improve the preparation quality and efficiency ,establish a sound whole-process cost estimate management system, and provide guidance, this standard has been developed according to the requirements of *Notice on Releasing the Development and Revision Plan of Energy Sector Standard in 2016* (GNKJ [2016] No. 238) issued by National Energy Administration of the People's Republic of China, after extensive investigation and research, summarization of practical experience, consultation of relevant standards of China, and wide solicitation of opinions.

Regulation for preparation of final account report for hydropower projects

1 Scope

This standard specifies the preparation method and composition of the final account report for hydropower projects.

This standard is applicable to the preparation of final account report for hydropower projects.

2 Normative references

The following referenced document is indispensable for the application of this document. For dated references, only the edition cited applies. For undated references, the latest edition of the referenced document (including any amendments) applies.

GB/T 14885, *Classification and codes for fixed assets*

Accounting Standards for Business Enterprises (Ministry of Finance of the People's Republic of China Order No. 33)

3 Terms and definitions

For the purpose of this document, the following terms and definitions apply.

3.1 approved budgetary estimate

design budgetary estimate of the hydropower project at feasibility study stage approved or authorized by the government

3.2 adjusted budgetary estimate

project investment document prepared for a hydropower project that has been approved and commenced in accordance with national and sector preparation regulations, in the case that the original approved budgetary estimate cannot meet the actual demand of the project due to government policy changes, price changes and design variations during the construction period

3.3 settlement at completion

document prepared by the employer and the contractor, on the basis of the adjusted and finalized contract prices, in accordance with the relevant national laws and regulations and contract agreement after the contractor has completed all the works as specified in the contract

3.4 final account

summary statement that reflects the construction costs, construction results and financial status, measured by physical quantities and monetary indicator

4 General provisions

4.1 This standard is developed in accordance with the relevant national regulations for hydropower projects to facilitate the preparation and management of final account report of hydropower projects and establish a sound hydropower project whole-process cost estimate management system, standardize the preparation principles, methods, content and level of detail for the final account report and improve the quality of preparation.

4.2 Only one final account report is prepared for the hydropower project(s) of an independent budget estimate.

4.3 The final account report is prepared by the project legal person, to ensure the authenticity, legality and integrity of the data in the final account report. And all the participants of the project such as the designer, supervision engineer, constructor and resettlement executor shall provide assistance in the preparation of the report.

4.4 The final account report is an important economic file, and shall be kept confidential and well custodies according to the relevant rules.

4.5 In addition to this standard, the preparation of the final account report shall comply with other current relevant standards of China.

5 Composition of final account report

5.1 Contents of final account report

The final account report consists of the cover, table of contents, photographs of the completed project, final account introduction, final account statements and associated important documents.

5.2 Cover of final account report

The cover of the final account report shall include the project name, project legal person, nature of the construction, final account base date, the date of preparation, legal representative, financial director etc., see Annex A for the format.

5.3 Photographs of the completed project

The photographs of the completed project shall reflect the panoramic view of project complex and the overview of the main structures, and the major resettlement areas, etc.

5.4 Final account introduction

The introduction summarizes the construction results and experiences of the completed project and analyze the investment performance of the project, which constitutes an important part of the final account report, including:

   a) Project overview.

   b) Preparation basis.

   c) Project prices settlement.

   d) Project cost accounting management.

   e) Project funds management.

   f) Basic construction procedure and management.

   g) Project budget estimates execution and analysis.

   h) Transfer of usable assets.

   i) Investment benefits.

   j) Other matters that need to be clarified.

   k) Records of major events.

5.5 Final account statements

The final account statements of hydropower projects, including:

   a) Fact sheet of the completed project.

   b) Final account sheet:

      1) Final account summary sheet.

2) Construction auxiliary works final account sheet.

3) Construction works final account sheet.

4) Environmental protection and soil and water conservation special works final account sheet.

5) Electro-mechanical equipment and installation works final account sheet.

6) Hydraulic steel structure and installation works final account sheet.

7) Resettlement compensation final account sheet.

8) Independent cost final account sheet.

c) Transfer of usable assets sheet:

1) Transfer of usable assets summary sheet.

2) Breakdown of transfer of usable fixed assets (structures and buildings).

3) Breakdown of transfer of usable fixed assets (equipment to be installed).

4) Breakdown of transfer of usable fixed assets (equipment not to be installed).

5) Breakdown of transfer of usable assets (current assets).

6) Breakdown of transfer of usable assets (intangible assets and long-term prepaid expenses).

d) Final financial account sheet.

e) Breakdown of yearly funds availability.

f) Breakdown of wind-up works and reserve fund.

g) Breakdown of prepaid expenses.

h) Prepaid expenses apportionment calculation sheet.

i) Breakdown of infrastructure expenditures to be written-off.

j) Breakdown of expenditures transferred-out.

See Annex B for the format of the final account statement for the completed hydropower projects.

5.6 Important documents of final account report

Important documents related of final account include:

a) Copies of project approvals, feasibility study reports and approval documents for budget estimates and budgetary estimate adjustment.

b) Copies of special items of approvals, acceptance and filing documents.

c) Copies of yearly expenditure plan and funds budget issuance documents.

d) Other documents related to the final account.

6 Final account report preparation
6.1 Preparation basis of final account report

The preparation basis of the final account report includes:

a) Relevant national laws, regulations and systems.

b) The approved feasibility study report, approved budget estimates and adjustment documents.

c) Annual investment plan and project expenditure budget.

d) Accounting and financial management data.

e) Tender documents, project contract or agreement and contract price adjustment documents.

f) Settlement at completion document.

g) Other relevant data.

6.2 Conditions for preparation of final account report

The conditions for preparation of the final account report shall include:

a) The project is completed and ready for operation or has passed the trial run.

b) Construction funds have been in place as planned.

c) The wind-up works and reserve fund do not exceed 5 % of the total investment in approved budget estimates or adjusted budget estimates.

d) The settlement at completion of the main works (other than the wind-up works) has been completed.

e) Issues involving lawsuit and project quality have been settled, and the resettlement compensation cost has been determined.

f) Other major issues affecting the preparation of the final account have been solved.

6.3 Preparation procedure of final account report

The procedure for the preparation of the final account report includes plan development, data collection, base date determination, financial liquidation, preparation of final account statements and final account report, etc.

6.4 Plan development of final account report

6.4.1 The project legal person shall formulate the plan development for final account report to give instruction and requirements for the preparation of the final account report.

6.4.2 The preparation plan should define the leadership and division of responsibilities, the base date of the final account, contents, schedule and procedures, technical problems and solutions, etc.

6.4.3 The responsibilities for the preparation of the final account report shall be broken down to departments and persons, and the work involving multiple departments shall be detailed and implemented in the plan.

6.5 Data collection of final account report

The data to be collected for the preparation of the final account report includes:

a) Approved budget estimates and adjustments.

b) Documents related to project design, preparation, approval, commencement and execution management.

c) Tender documents for the project, procurement of equipment and major materials and the implementation of relevant contracts, use of contingency.

d) Accounting vouchers, accounting books and accounting reports.

e) Internal management systems.

f) Documents related to the project prices settlement.

g) Documents of annual investment plan, budgetary estimates and execution status.

h) Approval documents for wind-up works and reserve fund.

i) The payment of resettlement compensation.

j) Bill of quantities of major construction works and list of material consumption.

k) The results of project acceptance and benefits.

l) Conclusive documents of audits, inspections, financial checks, and rectification.

m) Minutes of meetings, project briefing, work summary, etc.

6.6 Determination of final account base date

6.6.1 The base date of final account shall be determined considering comprehensively the project progress, the time of fund availability, settlement at completion and reimbursement, financial liquidation, etc.

6.6.2 The base date of final account should be the last day of the month.

6.6.3 The construction cost and asset value of the project shall enter into the account before the base date of final account.

6.7 Liquidation of final account

6.7.1 The final account financial liquidation includes contract liquidation, debt liquidation, fund balance liquidation and liquidation of transfer of usable assets.

6.7.2 The contract liquidation shall include:

a) List of contracts by numbering or category.

b) Settlement and payment of various contracts.

c) Confirmation of contract performance.

d) Correspondence between contract items and budget estimate items.

e) List of contracts remaining unclosed with outstanding payment amount, time limit of performance and measures.

6.7.3 The debt liquidation shall include:

a) Debt reconciliation.

b) Liquidation of the payables and receivables.

6.7.4 The fund balance liquidation shall include:

a) Verify the physical objects one by one, and make the list of physical objects that constitutes the balance fund.

b) Determine the handling methods and go through the related procedures.

6.7.5 The liquidation of transfer of usable assets shall include:

a) Make a book list of transfer of usable assets based on the accounting data.

b) Conduct on-site inventory check and generate the inventory list of the transfer of usable assets.

c) Compare and analyze the transfer of usable assets book list with the inventory list.

d) Adjust the discrepancies and generate a list of transfer of usable assets.

6.8 Preparation of final account statements

6.8.1 Principles for confirmation of final account statement values

The confirmation of the values of the final account statements shall follow the principles below:

a) The budgetary estimate value of the final account adopts the approved budgetary estimate value, and if there are adjustments, adopts the adjusted budgetary estimate value.

b) The actual value of the final account adopts the accounting book data, which shall include the wind-up works and reserve fund.

c) All the planned indicators adopt the plan data approved and issued by the superior department.

d) The actual techno-economic indicators and consumption of major materials shall adopt the real data of the project.

6.8.2 Preparation of fact sheet of the completed project

The fact sheet of the completed project reflects the scale, duration, investment, quality, techno-economic indicators, engineering characteristics, etc. of a hydropower project upon completion. The format is shown in Table B.1 of Annex B. The form shall be filled out following the guidance below:

a) The "nature of construction" refers to newly constructed, renovated or extended.

b) The "budgetary estimates approval department, document number and date" shall be as indicated in the final approval documents.

c) The "designer" refers to the designer of the main works. When there are two or more designers, only the responsible one shall be filled in.

d) The "supervisor" refers to the company undertaking the supervision of the main works.

e) The "construction contractor" refers to the company undertaking the construction of the main works. When two or more companies are involved, major construction companies may be filled in according to the actual situation.

f) The "main characteristics of project" is filled in according to the design report. If there are major variations in design during the construction, the actual main characteristics are filled in accordingly.

g) The "overall benefits" is filled in according to the design report. If major deviations occur during operation, the actual benefit is filled in.

h) The "sources of funds" include project capital, infrastructure appropriation, infrastructure loan and other funds. The project capital refers to the fund from the investors; the infrastructure appropriation refers to the financial appropriations from various financial departments; the infrastructure loan refers to the loans from banks or other financial institutions; and other funds refer to payables and non-payment. The actual amount of the above funds is filled in according to "sources of funds" in Table B.16 of Annex B.

i) The "actual investment" include transfer of usable assets, infrastructure expenditures to be written-off, transferred-out expenditures, input tax. The data of input tax shall be filled in according to "input tax" in Table B.2 of Annex B, and the rest of above shall be filled in according to Table B.16 of Annex B.

j) The items and amounts of "wind-up works and reserve fund" are filled in according to the corresponding amounts listed in the "total" under "estimated uncompleted investment" in Table B.18 of Annex B.

k) The "commencement date" is the commencement date of the first main works of construction installation works; the "completion date" is filled in according to the actual situations.

l) The "major works quantities" and "major material consumption" are filled in according to actual quantity within the scope of the main works.

m) The "overall quality evaluation" is filled in according to the comments from special acceptance committee.

6.8.3 Preparation of final account summary sheet

The final account summary sheet reflects the implementation of the budgetary estimates of the completed project. Table B.2 of Annex B is the summary sheet of Tables B.3 to B.9, and the formats are shown in Table B.2 to B.9 of Annex B. The items shall be consistent with the approved or adjusted budgetary estimates. The "item description" column shall be filled in following the sequence of the works listed in the approved or adjusted budgetary estimates.

a) Construction auxiliary works, construction works, environmental protection and soil and water conservation special works and independent costs may be grouped under class II items, and the equipment and installation works should be grouped under class III items. The formats are shown in Table B.3 to B.9 of Annex B.

b) In Table B.2 of Annex B, the estimated values of "basic contingency" and "contingency for price variation" are filled in according to the budgetary estimate, and the actual values are filled in according to the actual situations.

c) The input tax refers to the tax incurred during the project construction, which can be deducted by the project legal person. The "input tax" is filled in according to the accounting data.

6.8.4 Preparation of the transfer of usable assets summary sheet

The transfer of usable assets sheet reflects the composition of the assets transferred to production or receiving unit. It is the summary sheet of Table B.11 to B.15 in Annex B, and the format is shown in Table B.10 in Annex B. The tables shall be filled in as follows:

a) The "asset description" is classified as fixed assets, current assets, intangible assets and long-term prepaid expenses, which are listed in the order of the detailed breakdown of transfer of usable assets. The corresponding data shall be filled in according to the total values in the "asset value" in Table B.11 to B.15 of Annex B.

b) The fixed assets include structures, buildings, equipment to be installed and equipment not to be installed.

c) Current assets include the working capital and the stock materials, stock equipment, monetary funds and spare parts to be handed over to production.

d) Intangible assets include software, patents, copyrights, non-patented technology, trademark rights, land use right acquired through leasing, etc.

e) The long-term prepaid expenses include the production staff training cost and the production preparation cost in the nature of early mobilization fees.

f) The expenditures occurred in business-oriented projects that cannot form assets, such as river clearing and dredging, channel improvement, aerial seeding afforestation, cultivated land converted for forestry or pasture, hillside (desert) closure for afforestation, soil and water conservation, road repair, slope protection and clearing, etc., the expenditures of the unapproved project, and the expenses that happened before project cancellation and scrapping, shall be taken as prepaid expenses.

g) Dedicated associated facilities constructed for the project, including dedicated roads, communication facilities, power facilities, underground pipelines, dedicated docks, etc., the property rights of which do not belong to the company, shall be taken as intangible assets.

h) The lands that own the property transfer right shall be handled as intangible assets; the requisition expenses for the lands without land transfer right shall be taken as prepaid expenses.

i) For the auxiliary works or facilities constructed or procured for the project construction, including buildings, permanent roads, auxiliary roads, bridges, vehicles, office equipment, etc., accounting procedures shall be executed upon the completion of the project according to the following conditions. If the assets are directly delivered to the user, the self-used fixed assets that are depreciated during the project construction shall be delivered for use at the amount after deducting the accumulated depreciation from the acquisition cost of the fixed assets. The accrued depreciation expenses during the project construction shall be amortized to the relative asset values as prepaid expenses. If the assets are disposed of openly at varied prices before being delivered to the user, the accrued depreciation expenses during project construction and the net gain or loss (the difference between the openly sold-off price amount and the net asset prepaid expenses. For the items that the self-used fixed assets are not depreciated, the difference between the openly sold-off price value and the cost of acquisition shall be amortized to the value of the relative assets as the prepaid expenses.

6.8.4.1 Preparation of breakdown of transfer of usable fixed assets (structures and buildings)

The breakdown of transfer of usable fixed assets (structures and buildings) reflects the status of the structures and buildings that are delivered for production or to the receiving unit. The format is shown in Table B.11 of Annex B, and the form shall be filled out as follows:

a) The "item description" shall be filled in according to the registered object of assets directory in GB/T 14885. If the item cannot be found in the assets directory, the form shall be filled out following the provisions in *Accounting Standard for Business Enterprises*. If the building expenses can be recognized as individual fixed assets, this item shall be added; and if it cannot be recognized as a fixed asset, it shall be handled as prepaid expenses. The asset value of the transfer of usable fixed assets (structures and buildings) includes the construction costs and the prepaid expenses that need to be amortized but excludes the deductible input tax.

b) The "civil works cost" shall be filled in the order of "structures" and "buildings" in accordance with the actual accounting data, and the associated equipment in the structures and buildings such as lifts, central air-conditioners and others shall be handled as building construction expenses.

c) The "prepaid expenses" shall be filled in based on Table B.19 "breakdown of prepaid expenses" of Annex B, and the expenses shall be amortized to the construction costs according to the beneficiaries.

d) The "total" in the last row of the "civil works cost", "prepaid expenses" and "asset value" shall be filled in as "civil works cost" "prepaid expenses" and "total" items respectively corresponding to item "structures" and "buildings" in Table B.10 "transfer of usable assets summary sheet" of Annex B.

6.8.4.2 Preparation of breakdown of transfer of usable fixed assets (equipment to be installed)

The breakdown of transfer of usable fixed assets (equipment to be installed) reflects the status of the equipment to be delivered for production or to the receiving unit. The format is shown in Table 12 of Annex B.12. The equipment includes electro-mechanical equipment, hydraulic steel structures, etc., and the form shall be filled out as follows:

a) The "item description" shall be filled in according to the registered contents in GB/T 14885. Those that cannot be found in GB/T 14885 may be added to the list. The values of transfer of usable fixed assets (equipment to be installed) includes the procurement costs, installation costs and prepaid expenses of the equipment, and excludes the deductible input tax.

b) The "equipment procurement costs" shall be filled in as per the actual accounting data, and the costs include the original price of the equipment, transportation and miscellaneous costs, custody fee, etc., as well as customs duties, service charges for the imported equipment. The costs of tubes, earthing, cables (busbars), primary lines, rails, penstock, etc. in installation budgetary estimates shall also be listed in installation costs of final account. Cables (busbars), primary lines and penstock, shall be delivered as assets, and tubes, earthing, rails, etc. shall be included in the equipment associated but not delivered as independent assets.

c) The "installation works cost" shall be filled in as per the actual financial accounting data. The installation costs shall be listed firstly as works and shall be recorded separately if they are classified as fixed assets. For the installation costs that cannot be classified as fixed assets, such as heat preservation, painting, equipment lighting and metal testing costs, shall be amortized to the related works. When a works consists of different equipment, the installation cost in the works shall be amortized to different equipment, based on the actual equipment cost proportion.

d) The "total" value in the "equipment procurement costs", "installation works cost", "prepaid expenses" and "asset value" shall be filled in the corresponding columns "equipment procurement costs", "installation works cost", "prepaid expenses" and "total" in "equipment to be installed" in Table B.10 "transfer of usable assets summary sheet" of Annex B.

6.8.4.3 Preparation of breakdown of transfer of usable fixed assets (equipment not to be installed)

The breakdown of transfer of usable fixed assets (equipment not to be installed) reflects the status of equipment (not to be installed) transferred for production or to the receiving unit, and the format is shown in Table B.13 of Annex B. The equipment not to be installed mainly includes transportation vehicles, mechanical repair equipment, test equipment, tools, furniture, observation and monitoring equipment, etc. and the table shall be completed as below:

   a) The value of the equipment not to be installed shall be filled in according to the registration book kept by the asset management department and the relevant accounting data, excluding the deductible input tax.

   b) The amount of the "asset value" in "total" shall be filled in the corresponding column of "equipment not to be installed" in Table B.10 "transfer of usable assets summary sheet " of Annex B.

6.8.4.4 Preparation of breakdown of transfer of usable fixed assets (current assets)

The breakdown of transfer of usable fixed assets (current assets); reflects the status of the current assets transferred for production or to the receiving unit, the format is shown in Table B.14 of Annex B and the table shall be completed as below:

   a) The value of the current assets shall be filled in according to the registration book kept by asset management department and the relevant accounting data, excluding deductible input tax.

   b) The amount of "asset value" in "total" shall be filled in the corresponding column of "current assets" in Table B.10 "transfer of usable assets summary sheet " of Annex B.

6.8.4.5 Preparation of breakdown of transfer of usable fixed assets (intangible assets, long-term prepaid expenses)

The breakdown of transfer of usable fixed assets (intangible assets, long-term prepaid expenses) reflects the status of the intangible assets and long-term prepaid expenses delivered to the production unit, the format is shown in Table B.15 of Annex B, and the table shall be completed as below:

   a) The "item description" shall be filled in by the items of assets or costs, and the column "intangible assets" shall be filled in according to the relevant accounting data from the financial department, excluding deductible input tax.

   b) The "remarks" indicates the documents and agreements related to the intangible assets as well as the description of the assets delivery or funds transfer, for the convenience of assets management.

   c) The data for the "total" value of the "intangible assets" and "long-term prepaid expenses" shall be filled in the corresponding column of "intangible assets" and "long-term prepaid expenses" in Table B.10 "transfer of usable assets summary sheet" of Annex B.

6.8.5 Preparation of final financial account sheet

The final financial account sheet reflects the comprehensive situation of the sources and occupation of funds on the final account base date of the construction project, and the format is shown in Table B.16 of Annex B. The final financial account sheet shall adopt the balance sheet, i.e. all sources of funds shall be equal to all occupation of funds. The loan interest is included in the corresponding items, and the deposit interest of funds is included in offset of loan interest. The form shall be filled out as below:

   a) The items in "sources of funds" shall be filled in according to the "total" data of the corresponding items in Table B.17 of Annex B.

b) The items under "total transfer of usable assets" in column "occupation of funds" shall be filled in according to the corresponding data in column "asset value" in Table B.10 of Annex B.

c) The "infrastructure expenditures to be written-off" and "expenditures transferred-out" shall be filled in based on the total value of the "amount" column in Tables B.21 and B.22 of Annex B, respectively.

d) The "input tax" is the total value of the "input tax" column in Table B.2 of Annex B.

6.8.6 Preparation of breakdown of yearly funds availability

The breakdown of yearly funds availability reflects the annual funding status of hydropower projects, the format is shown in Table B.17 of Annex B. The sources of funds are listed as project capital, infrastructure appropriation, infrastructure loan and other funds. The project capital shall be listed by capital contribution ratio of the investors, capital nature and capital availability time. The infrastructure appropriation refers to the financial appropriations from various authorities and shall be listed by source and nature of funds. The infrastructure loans refer to the loans from the banks and shall be listed by the name of the bank, currency, loan term and interest rate. Other funds refer to the sources of funds other than the three listed above and normally refer to payables, non-payment, etc.

6.8.7 Preparation of breakdown of wind-up works and reserve fund

The breakdown of wind-up works and reserve fund reflects the remaining small amount of final works as well as the reserve fund such as acceptance costs after the base date of final account. The format is shown in Table B.18 of Annex B, and the form shall be filled out as below:

a) The "item description" shall be filled in according to the approved budgetary estimates, the construction works may be listed under the class-II items, and the equipment and installation works may be listed under class-III items.

b) The estimated wind-up works and reserve fund shall be filled in based on the planned expenses, and the completed works costs and the expected unfinished works costs shall be indicated.

6.8.8 Preparation of breakdown of prepaid expenses

Breakdown of prepaid expenses includes all the items under "prepaid expenses", the format is shown in Table B.19 of Annex B, and the form shall be filled out as below:

a) The "prepaid expenses" refer to the costs incurred from the construction within the approved project scope carried out by the project legal person, which shall be amortized and included in the related assets. It includes construction auxiliary works, river improvement works for navigation, slope protection and clearance work for near-dam slope treatment works, environmental protection and soil and water conservation works, compensation for resettlement, independent cost, interest during construction period and other costs.

b) The "other costs" include fixed assets losses, equipment disposal losses, equipment inventory losses and damage, scrapped works losses, damaged roads repair and other prepaid expenses.

c) The estimate value shall be the approved budgetary estimate and the actual data shall be filled in based on the financial accounting data. If the actual amount of the project includes the part that directly form the fixed assets, current assets, intangible assets or long-term prepaid expenses, the amount of this part shall be listed separately. The prepaid expenses shall be the balance after deducting the above-mentioned assets, and the nature of assets may refer to Table C.1 of Annex C.

6.8.9 Preparation of prepaid expenses apportionment calculation sheet

The prepaid expenses apportionment calculation sheet reflects the apportioning data of the actual amount of all the prepaid expenses, and the format is shown in Table B.20 of Annex B. The prepaid expenses may be apportioned step by step or in one step. Generally, the actual completed workloads is taken as the calculation basis for apportioning various costs, and if it is difficult to acquire the actual workloads, the estimated quantities may be adopted. Apportioning does not apply to the equipment not to be installed, current assets and intangible assets, and only the apportionment of the related compensation costs for resettlement applies to the corresponding structures and buildings.

a) When the step-by-step apportionment method is adopted, the beneficiaries of each apportioning expense shall be analyzed first, and the first round of apportioning shall refer to Table D.1 of Annex D. If this expense is only beneficial to a certain individual works, it may be apportioned directly into this individual works. If this expense is beneficial to either the structures and buildings works or the works of equipment to be installed, it may be apportioned to that works. If the expense is beneficial to both works above, the expenses shall be apportioned based on the percentage of the quantities of the two works, then the sum of the expenses in each works shall be apportioned to the breakdown items according to the proportions of the individual works.

b) When the one-step apportionment method is adopted, the actual (estimated) apportionment rate is calculated according to the total quantities of the construction works, installation works and the value of the equipment to be installed, the expenses shall be directly apportioned into the breakdown of each individual works.

c) The apportionment rate of prepaid expenses shall adopt the below formulae:

$$\text{Actual apportionment rate} = \frac{\text{expenses to be apportioned} - \text{expenses can be directly apportioned}}{\text{construction works balance} + \text{installation works balance} + \text{balance of equipment to be installed}}$$

$$\text{Budgetary estimate apportionment rate} = \frac{\text{prepaid expenses in budgetary estimate} - \text{expenses can be directly apportioned}}{\text{estimated construction works balance} + \text{estimated installation works balance} + \text{estimated balance of equipment to be installed}}$$

d) After the apportionment rate is obtained, the following formulae may be adopted to calculate the prepaid expenses to be apportioned to an asset:

$$\text{prepaid expenses of construction works} = \text{total construction works cost of the fixed asset} \times \text{apportionment rate}$$

$$\text{prepaid expenses of installation works} = \text{total installation works cost of the fixed asset} \times \text{apportionment rate}$$

prepaid expenses
of equipment to = total cost of equipment to be installed of the fixed asset ×apportionment rate
be installed

    e) The apportioning amount in each breakdown item shall be filled in the corresponding item in column "prepaid expenses" in Table B.11 and Table B.12 of Annex B.

6.8.10 Preparation of breakdown of infrastructure expenditures to be written-off

The breakdown of infrastructure expenditures to be written-off reflects the status of the construction expenditures, and the format is shown in Table B.21 of Annex B.

6.8.11 Preparation of breakdown of expenditures transferred-out

The breakdown of expenditures transferred-out reflects the status of the project expenditures transferred-out, and the format is shown in Table B.22 of Annex B.

6.8.12 Articulation among final account tables

6.8.12.1 Correlation between Table B.1 and Tables B.2 to B.22 in Annex B

    a) The total value of the "estimated costs" under "sources of funds" in Table B.1 is consistent with the total value of column 6 in Table B.2 of Annex B, and the total value of the "actual expenses" under "sources of funds" is equal to the total value of "sources of funds" in Table B.16 of Annex B.

    b) The "transfer of usable assets" under "actual costs" in Table B.1 is consistent with the total of "asset value" of column 7 in Table B.2 of Annex B.

    c) The "construction expenditures to be written-off", "expenditures transferred out" and "input tax" in Table B.1 shall be consistent with the total value of the corresponding items in Table B.16 of Annex B.

    d) The total value of "wind-up works and reserve fund" in Table B.1 is consistent with the total value of column 14 in Table B.18 of Annex B.

6.8.12.2 Correlation between Table B.2 and Tables B.10 to B.20 in Annex B

    a) The "estimated costs", "actual costs" and "difference between estimated costs and actual costs (including input tax)" columns of Table B.2 of Annex B corresponding to the items of "construction auxiliary works", "construction works", "environmental protection and soil and water conservation works" " electro-mechanical equipment and installation works", "compensation for resettlement" and "independent costs" shall be filled in according to the total values of corresponding items in Tables B.3 to B.9 of Annex B.

    b) The total value of column 11 in Table B.2 of Annex B is consistent with the total value of "asset value" of column 7 in Table B.10 of Annex B.

    c) The total value of column 13 in Table B.2 of Annex B is consistent with the total value of column "occupation of funds" in Table B.16 of Annex B.

    d) The "wind-up works and reserve fund" in Table B.2 of Annex B is consistent with the total values of column 14 in Table B.18 of Annex B.

6.8.12.3 Correlation between Table B.10 and Tables B.17 to B.20 in Annex B

    a) The "civil works cost", "prepaid expenses" and "asset value" under the items of "structures" and "buildings" in Table B.10 shall be filled in according to the total values of corresponding items in Table B.11 of Annex B.

  b) The "equipment procurement costs", "installation works cost", "prepaid expenses" and "asset value" under "equipment to be installed" in Table B.10 of Annex B shall be filled in according to the total values of corresponding items in Table B.12 of Annex B.

  c) The "equipment procurement costs" under "equipment not to be installed" in Table B.10 of Annex B shall be filled in according to the total value of column "asset value" in Table B.13 of Annex B.

  d) The "expenses directly identified as assets under "current assets" in Table B.10 of Annex B shall be filled in according to the total value of column "asset value" in Table B.14 of Annex B.

  e) The "expenses directly identified as assets" under items of "intangible assets" and "prepaid expenses" in Table B.10 of Annex B shall be filled in according to the total value of column "asset value" in Table B.15 of Annex B.

  f) The "asset value" under items "fixed assets", "current assets", "intangible assets" and "long-term prepaid expenses" in Table B.10 of Annex B shall be filled in according to the total values of corresponding items in Table B.16 of Annex B.

  g) The "prepaid expenses" in Table B.10 of Annex B is consistent with the total values of column 3 in Table B.19 of Annex B.

  h) The "prepaid expenses" in Table B.10 of Annex B is consistent with the total value of column "prepaid expenses" in Table B.20 of Annex B; The "prepaid expenses" in Table B.11 and B12 of Annex B shall be filled in according to Table B.20 of Annex B.

6.8.12.4 Correlation between Table B.16 and Tables B.18 to B.20 of Annex B

  a) The total value of "sources of funds" in Table B.16 of Annex B is consistent with the total value of column "total" in Table B.17 of Annex B.

  b) There are no correlations between Table B.16 and Table B.19 and Table B.20 of Annex B.

  c) The data of "construction expenditures to be written-off" and "expenditures transferred out" in Table B.16 of Annex B shall be consistent with the total values in Table B.21 and B.22 of Annex B.

6.8.12.5 Correlation between Table B.19 and Table B.20 of Annex B

Table B.20 is the attachment of B.19, and the total value of "prepaid expenses" in Table B.19 of Annex B is consistent with the total value of "prepaid expenses" in Table.20 of Annex B.

6.9 Preparation of final account introduction

The final account introduction shall be prepared as follows:

  a) Project overview: mainly state the project objectives and benefits, the purpose of construction, project history, project legal person, project design process, construction process of main works, compensation for resettlement, etc.

  b) Preparation basis and method: introduce the basis, principles, methods, composition and breakdown of the works.

  c) Project contract price and payment: mainly discuss the signing of contractual agreement and payments status, the comparison between the project costs and the costs of similar projects of the same period, as well as the status of the wind-up works and the reserve fund.

d) Project cost accounting management: mainly state the construction costs accounting method, including the expenditures beyond the scope of the project, expenditures not complying with contractual agreements, expenditures with incomplete invoices or without invoices, expenditures without approval procedures or signatures of responsible persons, and the status of scraping loss and the expenditures not belonging to the project, etc. It describes the status of accounting transactions, property and materials liquidation, debts liquidation, prepaid expenses apportionment, expenditures on infrastructure to be written off and expenditures to be transferred out.

e) Project funds management: mainly describe the conditions of the project construction funding, funding costs control, project budget estimate, plan, funds availability and use, the allocation of income and balance funds.

f) Capital construction procedure implementation and management: mainly describe the project decision-making process, pre-feasibility study, feasibility study, approval, major design variation and budgetary estimate adjustments, project design management, construction management, tender and equipment procurement management, planning and contract management, rationalization suggestions, application of advanced scientific methods and technologies, lessons from experience, etc.

g) Project budget estimate implementation and analysis: mainly discuss whether the project is implemented in accordance with the approved budget estimate, whether there are works exceeding standards, scale or budget estimate, the rectification and implementation of the comments from each audit, check, inspection and review, the status of wind-up works and reserve fund, analysis of the differences between actual expenditures and budget estimate, as well as the use of the contingency.

h) Transfer of usable assets status: mainly describe project assets identification, whether the actual cost of the transfer of usable assets is complete, whether the conditions of transfer are met, and the receiving unit of the assets.

i) Investment benefit status: mainly describe the analysis and calculation of the main techno-economic indicators such as the project payback period (static and dynamic), financial net present value and internal rate of return, the analysis of the loan repayment capacity, and the evaluation of the economic, social and environmental benefits of the project.

j) Others: describe the treatment and methods used in the preparation that have significant impacts on the data.

k) Record of major events: record the relevant activities during preparation and construction as well as the milestones and other significant matters relating to project management, which shall be listed in chronological order.

Annex A
(Normative)
Format of the cover of final account report for hydropower projects

Figure A.1 specifies the content and format of the cover of final account report for a hydropower project.

CONFIDENTIAL

Final Account Report of Hydropower Project

Project name:_____

Project legal person:_____

Nature of construction:_____

Base date of final account:_____

Preparation date:_____

Legal representative:_____   Financial director:_____

Figure A.1   Format of the cover of final account report

Annex B
(Normative)

Format of final account statement sheets for hydropower projects

Tables B.1 to B.22 specify the format of final account statement sheets for hydropower projects.

Table B.1 Fact sheet of the completed project

Completed project-Table 01

Project name			Nature of construction			Location and river		
Project legal person			Designer			Construction contractor		
Supervisor			Quality supervisor					
Budgetary estimate approval department, document number and date								
main characteristics of project				Sources of funds			Actual investment	
				Item	Budgetary estimate	Actual fund	Ⅰ. Total transfer of usable assets	
Reservoir	Reservoir name			1. Project capital			Fixed assets	
	Normal pool level (m)		Project Investment (CNY)	2. Infrastructure appropriation			Current assets	
	Total storage (m³)			3. Infrastructure loan			Intangible assets	
	Effective storage (m³)			4. Other funds			Long-term prepaid expenses	
Dam	dam type			…			Ⅱ. Construction expenditures to be written-off	
	Maximum height/crest Length (m)			Total			Ⅲ. Expenditures transferred out	
Spillway	Type		Wind-up works and reserve fund	Work description			Ⅳ. Input tax	
	Number and size of bays/outlets						Estimated investment	
	Maximum discharge (m³/s)							
Powerhouse	type and size (l×h×w)							
	Installed capacity (single unit capacity × number of units)							

Completed project -Table 01

Table B.1 Fact sheet of the completed project *(continued)*

				Project construction overview			
	Firm output (MW)				Item	Design (plan)	Actual
	Annual energy output (kW·h)				Commencement date		
	Equipment description		Specification/model		Completion date		
Major equipment	Turbine			Major works quantities	1. Earth and rock excavation (m³)		
	Generator				2. Earth and rock fill (m³)		
	Main transformer				3. Concrete (m³)		
Comprehensive benefits	1. Flood control				4. Hydraulic steel structure fabrication and installation (t)		
	2. Irrigation				...		
	3. Water supply			Major material consumption	1. Rebar/steel (t)		
	Item	Design (plan)	Actual		2. Cement(t)		
Resettlement	1. Total compensation fee (CNY)				3. Fuels (t)		
	2. Land requisition (acre)				...		
	3. Relocatee (person)			Overall quality evaluation			
	4. House demolition (m²)						
	5. Per capita compensation fee (CNY/person)						

NOTE   The major characteristics of the reservoirs, dams and spillways of a pumped-storage power station shall be described separately by the upper and lower reservoirs.

Completed project – Table 02

Table B.2 Final account summary sheet

Unit: CNY

No.	Item description	Estimated costs					Actual costs						Difference between actual and estimated costs	
		Civil	Installation	Equipment procurement	Others	Total	Civil	Installation	Equipment procurement	Others	Subtotal	Input tax	Total	Increase / decrease	Increase / decrease rate (%)
	1	2	3	4	5	6=2+3+4+5	7	8	9	10	11=7+8+9+10	12	13=11+12	14=13−6	15=14/6
I	Project complex														
1	Construction auxiliary works														
2	Civil works														
3	Environmental protection and soil and water conservation works														
4	Electro-mechanical equipment and installation works														
5	Hydraulic steel structure equipment and installation works														
II	Resettlement compensation														
III	Independent cost														

Completed project – Table 02

Table B.2 Final account summary sheet *(continued)*

Unit: CNY

No.	Item description	Estimated costs					Actual costs						Difference between actual and estimated costs	
		Civil	Installation	Equipment procurement	Others	Total	Civil	Installation	Equipment procurement	Others	Subtotal	Input tax	Total	Increase / decrease	Increase / decrease rate (%)
	1	2	3	4	5	6=2+3+4+5	7	8	9	10	11=7+8+9+10	12	13=11+12	14=13-6	15=14/6
1	Project construction management cost														
2	Production preparation fee														
3	Research, investigation and design cost														
4	Other taxes and fees														
IV	Basic contingency														
V	Contingency for price variation														
VI	Interest during construction period														
	Total														
	Of which: wind-up works and reserve fund														

Completed project – Table 02-1

Table B.3　Construction auxiliary works final account sheet

Unit: CNY

No.	Item description	Estimated costs					Actual costs						Difference between actual and estimated costs	
		Civil	Installation	Equipment procurement	Others	Total	Civil	Installation	Equipment procurement	Others	Subtotal	Input tax	Total	Increase / decrease	Increase / decrease rate (%)
1		2	3	4	5	6=2+3+4+5	7	8	9	10	11=7+8+9+10	12	13=11+12	14=13-6	15=14/6
	Construction auxiliary works														
I	Construction transportation														
...															
II	Navigation during construction														
...															
III	Construction power supply														
...															
IV	Construction water supply														
...															
V	Construction compressed-air supply														
...															
VI	Construction communications														
...															
VII	Construction management information system														
...															
VIII	Quarry/borrow area overburden stripping and protection														
...															

Completed project – Table 02-1

Table B.3 Construction auxiliary works final account sheet *(continued)*

Unit: CNY

No.	Item description	Estimated costs					Actual costs						Difference between actual and estimated costs	
		Civil	Installation	Equipment procurement	Others	Total	Civil	Installation	Equipment procurement	Others	Subtotal	Input tax	Total	Increase / decrease	Increase / decrease rate (%)
	1	2	3	4	5	6=2+3+4+5	7	8	9	10	11=7+8+9+10	12	13=11+12	14=13-6	15=14/6
IX	Aggregate processing system														
...															
X	Concrete production and pouring system														
...															
XI	Diversion works														
...															
XII	Temporary safety monitoring works														
...															
XIII	Hydrologic telemetry and forecasting works														
...															
XIV	Construction and management building works														
...															
XV	Other construction auxiliary works														
...															
	Total														
	Of which: wind-up works and reserve fund														

NB/T 10145-2019

Completed project – Table 02- 2

Table B.4 Construction works final account sheet

Unit: CNY

| No. | Item description | Estimated costs ||||| Actual costs ||||| Difference between actual and estimated costs ||
		Civil	Installation	Equipment procurement	Others	Total	Civil	Installation	Equipment procurement	Others	Subtotal	Input tax	Total	Increase / decrease	Increase / decrease rate (%)
	1	2	3	4	5	6=2+3+4+5	7	8	9	10	11=7+8+9+10	12	13=11+12	14=13-6	15=14/6
	Construction works														
I	Water retaining (storage) structure														
...															
II	Discharge and energy dissipation structure														
...															
III	Water conveyance structure														
...															
IV	Power generation structure														
...															
V	Step-up substation structure														
...															

Completed project – Table 02-2

Table B.4 Construction works final account sheet *(continued)*

Unit: CNY

No.	Item description	Estimated costs					Actual costs						Difference between actual and estimated costs	
		Civil	Installation	Equipment procurement	Others	Total	Civil	Installation	Equipment procurement	Others	Subtotal	Input tax	Total	Increase / decrease	Increase / decrease rate (%)
	1	2	3	4	5	6=2+3+4+5	7	8	9	10	11=7+8+9+10	12	13=11+12	14=13−6	15=14/6
VI	Dam-passing navigation structure														
...															
VII	Irrigation canal head structure														
...															
VIII	Abutment slope treatment works														
...															
IX	Transport works														
...															
X	Building works														
...															
XI	Safety monitoring works														
XII	Hydrologic telemetry and forecasting works														

NB/T 10145-2019

Completed project – Table 02-2

Table B.4 Construction works final account sheet *(continued)*

Unit: CNY

No.	Item description	Estimated costs					Actual costs						Difference between actual and estimated costs	
		Civil	Installation	Equipment procurement	Others	Total	Civil	Installation	Equipment procurement	Others	Subtotal	Input tax	Total	Increase / decrease	Increase / decrease rate (%)
1	1	2	3	4	5	6=2+3+4+5	7	8	9	10	11=7+8+9+10	12	13=11+12	14=13-6	15=14/6
XIII	Fire protection works														
XIV	Occupational health and safety works														
XV	Other works														
1	Power line works														
2	Lighting line works														
3	Communication line works														
4	Water supply and drainage works in the powerhouse and dam areas														
...															
	Total														
	Of which: wind-up works and reserve fund														

25

Completed project – Table 02-3
Unit: CNY

Table B.5 Environmental protection and soil and water conservation special works final account sheet

No.	Item description	Estimated costs					Actual costs						Difference between actual and estimated costs	
		Civil	Installation	Equipment procurement	Others	Total	Civil	Installation	Equipment procurement	Others	Subtotal	Input tax	Total	Increase / decrease	Increase / decrease rate (%)
1	1	2	3	4	5	6=2+3+4+5	7	8	9	10	11=7+8+9+10	12	13=11+12	14 = 13-6	15=14/6
	Environmental protection and soil and water conservation works														
I	Environmental protection														
1	Water environment protection														
2	Atmospheric environment protection														
3	Acoustic environment protection														
4	Solid waste disposal														
5	Soil environment protection														
6	Terrestrial ecosystem protection														
7	Aquatic ecosystem protection														
8	Population health protection measures														

NB/T 10145-2019

Completed project – Table 02-3

Table B.5 Environmental protection and soil and water conservation special works final account sheet *(continued)*

Unit: CNY

No.	Item description	Estimated costs					Actual costs					Difference between actual and estimated costs	
		Civil	Installation	Equipment procurement	Others	Total	Civil	Installation	Equipment procurement	Others	Subtotal	Input tax	Total	Increase / decrease	Increase / decrease rate (%)
	1	2	3	4	5	6=2+3+4+5	7	8	9	10	11=7+8+9+10	12	13=11+12	14 = 13–6	15=14/6
9	Landscape conservation works														
10	Environmental monitoring (survey)														
11	Other environmental protection works														
II	Soil and water conservation works														
1	Engineering measures														
2	Vegetation measures														
3	Soil and water conservation monitoring works														
4	Others														
	Total														
	Of which: wind-up works and reserve fund														

27

Completed project – Table 02-4

Table B.6 Electro-mechanical equipment and installation works final account sheet

Unit: CNY

No.	Item description	Estimated costs			Actual costs										Difference between actual and estimated costs	
		Installation	Equipment procurement	Total	Installation				Equipment procurement				Total			Increase / decrease	Increase / decrease rate (%)
					Input tax excl.	Input tax	Input tax incl.	Difference between actual & estimated costs	Input tax excl.	Input tax	Input tax incl.	Difference between actual & estimated costs	Input tax excl.	Input tax	Input tax incl.		
	1	2	3	4=2+3	5	6	7=5+6	8=7−2	9	10	11=9+10	12=11−3	13=5+9	14=6+10	15=7+11	16=15−4	17=16/4
	Electro-mechanical equipment and installation works																
I	Power generation equipment and installation																
1	Turbine (pump-turbine) equipment and installation																
...																	
2	Generator (generator-motor) equipment and installation																
...																	
3	Inlet valve equipment and installation																
...																	

NB/T 10145-2019

Completed project – Table 02-4
Unit: CNY

Table B.6 Electro-mechanical equipment and installation works final account sheet *(continued)*

No.	Item description	Estimated costs			Actual costs									Difference between actual and estimated costs	
		Installation	Equipment procurement	Total	Installation				Equipment procurement				Total		
					Input tax excl.	Input tax	Input tax incl.	Difference between actual & estimated costs	Input tax excl.	Input tax	Input tax incl.	Difference between actual & estimated costs	Input tax excl.	Input tax	Input tax incl.	Increase / decrease	Increase / decrease rate (%)
1		2	3	4=2+3	5	6	7=5+6	8=7-2	9	10	11=9+10	12=11-3	13=5+9	14=6+10	15=7+11	16=15-4	17=16/4
4	Hoist equipment and installation																
...																	
5	Hydraulic machinery auxiliary equipment and installation																
...																	
6	Electrical equipment and installation																
...																	
7	Control and protection equipment and installation																
...																	
8	Communication equipment and installation																

Completed project – Table 02-4

Unit: CNY

Table B.6 Electro-mechanical equipment and installation works final account sheet *(continued)*

No.	Item description	Estimated costs			Actual costs									Difference between actual and estimated costs	
		Installation	Equipment procurement	Total	Installation				Equipment procurement				Total		Increase / decrease	Increase / decrease rate (%)
					Input tax excl.	Input tax	Input tax incl.	Difference between actual & estimated costs	Input tax excl.	Input tax	Input tax incl.	Difference between actual & estimated costs	Input tax excl.	Input tax	Input tax incl.		
	1	2	3	4=2+3	5	6	7=5+6	8=7−2	9	10	11=9+10	12=11−3	13=5+9	14=6+10	15=7+11	16=15−4	17=16/4
…																	
II	Step-up substation equipment and installation																
1	Main transformer equipment and installation																
…																	
2	High voltage electrical equipment and installation																
…																	
3	Primary circuit line and other installation works																
III	Dam-passing navigation equipment and installation																

NB/T 10145-2019

Completed project – Table 02-4

Table B.6 Electro-mechanical equipment and installation works final account sheet *(continued)*

Unit: CNY

No.	Item description	Estimated costs			Actual costs										Difference between actual and estimated costs	
		Installation	Equipment procurement	Total	Installation				Equipment procurement				Total			Increase / decrease	Increase / decrease rate (%)
					Input tax excl.	Input tax	Input tax incl.	Difference between actual & estimated costs	Input tax excl.	Input tax	Input tax incl.	Difference between actual & estimated costs	Input tax excl.	Input tax	Input tax incl.		
	1	2	3	4=2+3	5	6	7=5+6	8=7-2	9	10	11=9+10	12=11-3	13=5+9	14=6+10	15=7+11	16=15-4	17=16/4
1	Power supply equipment and installation																
2	Control equipment and installation																
IV	Safety monitoring equipment and installation																
V	Hydrologic telemetry and forecasting equipment and installation																
VI	Fire protection equipment and installation																

31

Completed project – Table 02-4

Table B.6 Electro-mechanical equipment and installation works final account sheet *(continued)*

Unit: CNY

No.	Item description	Estimated costs			Actual costs										Difference between actual and estimated costs	
		Installation	Equipment procurement	Total	Installation				Equipment procurement				Total			
					Input tax excl.	Input tax	Input tax incl.	Difference between actual & estimated costs	Input tax excl.	Input tax	Input tax incl.	Difference between actual & estimated costs	Input tax excl.	Input tax	Input tax incl.	Increase / decrease	Increase / decrease rate (%)
	1	2	3	4=2+3	5	6	7=5+6	8=7−2	9	10	11=9+10	12=11−3	13=5+9	14=6+10	15=7+11	16=15−4	17=16/4
VII	Occupational health and safety equipment and installation																
VIII	Other equipment and installation																
1	Lifts and installation																
…																	
2	Feeder equipment and installation in dam area																
…																	
3	Water supply and drainage equipment and installation in powerhouse and dam areas																

NB/T 10145-2019

Completed project – Table 02-4

Table B.6 Electro-mechanical equipment and installation works final account sheet *(continued)*

Unit: CNY

No.	Item description	Estimated costs			Actual costs										Difference between actual and estimated costs	
		Installation	Equipment procurement	Total	Installation				Equipment procurement				Total			
					Input tax excl.	Input tax	Input tax incl.	Difference between actual & estimated costs	Input tax excl.	Input tax	Input tax incl.	Difference between actual & estimated costs	Input tax excl.	Input tax	Input tax incl.	Increase / decrease	Increase / decrease rate (%)
1		2	3	4=2+3	5	6	7=5+6	8=7-2	9	10	11=9+10	12=11-3	13=5+9	14=6+10	15=7+11	16=15-4	17=16/4
4	Heat supply equipment and installation in powerhouse and dam areas																
5	Cascade centralized control center equipment costs apportionment																
6	Ventilation and heating equipment and installation																
7	Mechanical repair equipment and installation																

33

NB/T 10145-2019

Completed project – Table 02-4

Table B.6 Electro-mechanical equipment and installation works final account sheet *(continued)*

Unit: CNY

No.	Item description	Estimated costs			Actual costs											Difference between actual and estimated costs	
		Installation	Equipment procurement	Total	Installation				Equipment procurement				Total			Increase / decrease	Increase / decrease rate (%)
					Input tax excl.	Input tax	Input tax incl.	Difference between actual & estimated costs	Input tax excl.	Input tax	Input tax incl.	Difference between actual & estimated costs	Input tax excl.	Input tax	Input tax incl.		
1		2	3	4=2+3	5	6	7=5+6	8=7−2	9	10	11=9+10	12=11−3	13=5+9	14=6+10	15=7+11	16=15−4	17=16/4
8	Earthquake monitoring station network equipment																
9	Transport vehicles																
10	Plant area earthing																
11	Others																
…																	
	Total																
	Of which: wind-up works and reserve fund																

NB/T 10145-2019

Completed project – Table 02-5

Table B.7 Hydraulic steel structures and installation works final account sheet

Unit: CNY

No.	Item description	Estimated costs			Actual costs										Difference between actual and estimated costs	
		Installation	Equipment procurement	Total	Installation				Equipment procurement				Total			
					Input tax excl.	Input tax	Input tax incl.	Difference between actual & estimated costs	Input tax excl.	Input tax	Input tax incl.	Difference between actual & estimated costs	Input tax	Input tax incl.	Difference between actual & estimated costs	Increase / decrease	Increase / decrease rate (%)
	1	2	3	4=2+3	5	6	7=5+6	8=7-2	9	10	11=9+10	12 = 11-3	13=5+9	14=6+10	15=7+11	16 = 15-4	17 = 16/4
	Hydraulic steel structures and installation works																
I	Water retaining (storage) structure																
1	Gates and installation																
...																	
2	Gate hoists and installation																
...																	
3	Trash racks and installation																
...																	
II	Discharge and energy dissipation structure																

Completed project – Table 02-5

Table B.7 Hydraulic steel structures and installation works final account sheet *(continued)*

Unit: CNY

No.	Item description	Estimated costs			Actual costs										Difference between actual and estimated costs	
		Installation	Equipment procurement	Total	Installation				Equipment procurement				Total		Increase / decrease	Increase / decrease rate (%)
					Input tax excl.	Input tax	Input tax incl.	Difference between actual & estimated costs	Input tax excl.	Input tax	Input tax incl.	Difference between actual & estimated costs	Input tax	Input tax incl.	Difference between actual & estimated costs		
	1	2	3	4=2+3	5	6	7=5+6	8=7−2	9	10	11=9+10	12=11−3	13=5+9	14=6+10	15=7+11	16=15−4	17=16/4
1	Gates and installation																
2	Gate hoists and installation																
3	Trash racks and installation																
III	Water conveyance structure																
1	Gates and installation																
2	Gate hoists and installation																
3	Trash racks and installation																
4	Steel penstock fabrication and installation																

NB/T 10145-2019

Completed project – Table 02-5

Table B.7 Hydraulic steel structures and installation works final account sheet (continued)

Unit: CNY

No.	Item description	Estimated costs			Actual costs										Difference between actual and estimated costs	
		Installation	Equipment procurement	Total	Installation				Equipment procurement				Total		Increase / decrease	Increase / decrease rate (%)
					Input tax excl.	Input tax	Input tax incl.	Difference between actual & estimated costs	Input tax excl.	Input tax	Input tax incl.	Difference between actual & estimated costs	Input tax	Input tax incl.	Difference between actual & estimated costs		
	1	2	3	4=2+3	5	6	7=5+6	8=7−2	9	10	11=9+10	12 = 11−3	13=5+9	14=6+10	15=7+11	16 = 15−4	17 = 16/4
IV	Step-up substation structure																
1	Steel members																
V	Dam-passing navigation structure																
1	Gates and installation																
2	Gate hoists and installation																
3	Shiplift and installation																
4	Dam-passing equipment and installation																
...																	

NB/T 10145-2019

Completed project – Table 02-5

Unit: CNY

Table B.7 Hydraulic steel structures and installation works final account sheet *(continued)*

No.	Item description	Estimated costs			Actual costs										Difference between actual and estimated costs	
		Installation	Equipment procurement	Total	Installation				Equipment procurement			Total			Increase / decrease	Increase / decrease rate (%)
					Input tax excl.	Input tax	Input tax incl.	Difference between actual & estimated costs	Input tax excl.	Input tax	Input tax incl.	Difference between actual & estimated costs	Input tax	Input tax incl.	Difference between actual & estimated costs		
	1	2	3	4=2+3	5	6	7=5+6	8=7-2	9	10	11=9+10	12=11-3	13=5+9	14=6+10	15=7+11	16=15-4	17=16/4
VI	Irrigation canal head structure																
1	Gates and installation																
2	Gate hoists and installation																
...																	
	Total																
	Of which: wind-up works and reserve fund																

Completed project – Table 02-6

Table B.8 Resettlement compensation final account sheet

Unit: CNY

No.	Item description	Estimated costs					Actual costs						Difference between actual and estimated costs	
		Civil	Installation	Equipment procurement	Others	Subtotal	Civil	Installation	Equipment procurement	Others	Subtotal	Input tax	Total	Increase / decrease	Increase / decrease
1	1	2	3	4	5	6=2+3+4+5	7	8	9	10	11=7+8+9+10	12	13=11+12	14 = 13-6	15=14/6
	Resettlement compensation														
I	Reservoir Impoundment-affected area														
1	Rural area														
2	Town area														
3	Special items														
4	Reservoir basin clearance														
5	Environmental protection and soil and water conservation														
II	Project complex area														
1	Rural area														
2	Town area														
3	Special items														
4	Reservoir basin clearance														
5	Environmental protection and soil and water conservation														
	Total														
	Of which: wind-up works and reserve fund														

NB/T 10145-2019

Completed project – Table 02-7

Table B.9 Independent cost final account sheet

Unit: CNY

No.	Item description	Estimated costs				Actual costs				Difference between actual and estimated costs	
		Civil	Installation	Equipment procurement	Civil	Installation	Equipment procurement	Civil	Installation	Civil	Installation
1	1	2	3	4	5	6=2+3+4+5	7	8	9	10	11=7+8+9+10	12	13=11+12	14=13-6	15=14/6
I	Independent cost										
1	Project construction management cost										
2	Preliminary engineering fee										
3	Owner's construction management cost										
4	Resettlement compensation management cost										
5	Construction supervision cost										
6	Resettlement monitoring and assessment cost										
7	Consulting service cost										
	Technical and economic evaluation cost										

Table B.9 Independent cost final account sheet *(continued)*

Completed project – Table 02-7

Unit: CNY

No.	Item description	Estimated costs					Actual costs					Difference between actual and estimated costs	
		Civil	Installation	Equipment procurement	Civil	Installation	Equipment procurement	Civil	Installation	Equipment procurement	Civil	Civil	Installation
	1	2	3	4	5	6=2+3+4+5	7	8	9	10	11=7+8+9+10	12	13=11+12	14 = 13−6	15=14/6
8	Cost for project quality inspection and test												
9	Management cost for norm standard preparation for hydropower project												
10	Project acceptance cost												
11	Construction insurance fee												
II	Production preparation fee												
III	Research, investigation and design cost												
1	Construction research and test cost												
2	Investigation and design cost												

NB/T 10145-2019

Completed project – Table 02-7

Table B.9 Independent cost final account sheet *(continued)*

Unit: CNY

No.	Item description	Estimated costs					Actual costs					Difference between actual and estimated costs	
		Civil	Installation	Equipment procurement	Civil	Installation	Equipment procurement	Civil	Installation	Equipment procurement	Civil	Installation	
	1	2	3	4	5	6=2+3+4+5	7	8	9	10	11=7+8+9+10	12	13=11+12	14 = 13-6	15=14/6
IV	Other taxes and fees													
1	Arable land occupation tax													
2	Arable land reclamation cost													
3	Forest and vegetation restoration cost													
4	Compensation fee for soil and water conservation facilities													
5	Others													
	Total													
	Of which: wind-up works and reserve fund													

Completed project – Table 03

Table B.10 Transfer of usable assets summary sheet

Unit: CNY

No.	Asset description	Civil works cost	Installation works cost	Equipment procurement cost	Other costs		Asset value	Remarks
					Prepaid expenses	Expenses directly identified as assets		
	1	2	3	4	5	6	7=2+3+4+5+6	8
I	Fixed assets							
	Structures							
	Buildings							
	Equipment to be installed							
	Equipment not to be installed							
II	Current assets							
III	Intangible assets							
IV	Long-term prepaid expenses							
	Total							

NB/T 10145-2019

NB/T 10145-2019

Completed project – Table 03-1

Table B.11 Breakdown of transfer of usable fixed assets (structures and buildings)

Unit: CNY

No.	Item description	Specifications, models, features	Location	Unit of measurement	Quantity	Civil works cost	Prepaid expenses	Asset value	Remarks
1	2	3	4	5	6	7	8=6+7	9	
I	Structures								
(I)	Water retaining (storage) works								
1	Concrete dam	Dam type, maximum dam height/ dam axis length							
2	Earth/Rock fill dam	Dam type, maximum dam height/dam axis length							
	...								
(II)	Water release and energy dissipation structure								
1	Spillway	Type, hole size							
2	Flood discharge tunnel	Diameter and length							
	...								
(III)	Water conveyance structure								
1	Headrace tunnel	Diameter and length							
	...								
(IV)	Power generation structure								
1	Surface powerhouse	Length × Width × Height							
2	Underground powerhouse	Length × Width × Height							
	...								
(V)	Step-up substation structure								

NB/T 10145-2019

Table B.11 Breakdown of transfer of usable fixed assets (structures and buildings) (continued)

Completed project – Table 03-1

Unit: CNY

No.	Item description	Specifications, models, features	Location	Unit of measurement	Quantity	Civil works cost	Prepaid expenses	Asset value	Remarks
1	2	3	4	5	6	7	8=6+7	9
1	Surface substation	Voltage level, dimension							
...									
(VI)	Dam-passing navigation structure								
1	Upstream approach channel works								
2	Lock(shiplift) works	Grade, size							
...									
(VII)	Irrigation headworks	Type, orifice size							
(VIII)	Abutment slope treatment works	Length							
(IX)	Transport works								
1	Highway works	Grade, length							
...									
(X)	Safety monitoring works								
(XI)	Hydrologic telemetry and forecasting works								
(XII)	Fire protection								
(XIII)	Occupational health and safety								
(XIV)	Other works								
1	Power line works	Grade, length							
2	Lighting line works	Length							

NB/T 10145-2019

Table B.11　Breakdown of transfer of usable fixed assets (structures and buildings) *(continued)*

Completed project – Table 03-1

Unit: CNY

No.	Item description	Specifications, models, features	Location	Unit of measurement	Quantity	Civil works cost	Prepaid expenses	Asset value	Remarks
1	2	3	4	5	6	7	8=6+7	9	
3	Communication line works								
4	Water supply and drainage works in the powerhouse and dam areas								
5	Heat supply works in the powerhouse and dam areas								
6	Earthquake monitoring station network								
7	Others								
...									
(XV)	Environmental protection works								
...									
(XVI)	Soil and water conservation works								
...									
II	Buildings								
1	Auxiliary production plant								
2	Warehouses								

NB/T 10145-2019

Table B.11 Breakdown of transfer of usable fixed assets (structures and buildings) *(continued)*

Completed project – Table 03-1

Unit: CNY

No.	Item description	Specifications, models, features	Location	Unit of measurement	Quantity	Civil works cost	Prepaid expenses	Asset value	Remarks
1	2	3	4	5	6	7	8=6+7	9
3	Office buildings								
4	On-duty flat and ancillary facilities								
5	Production and operation management facilities								
6	Apportioning cost of central control center of cascade power stations								
	Total								

47

NB/T 10145-2019

Completed project – Table 03-2

Table B.12 Breakdown of transfer of usable fixed assets (equipment to be installed)

Unit: CNY

No.	Item description	Specification/ model	Supplier/ manufacturer	Installation location	Unit of measurement	Quantity	Equipment procurement cost	Installation works cost	Prepaid expenses	Asset value	Remarks
	1	2	3	4	5	6	7	8	9	10=7+8+9	11
I	Electro-mechanical equipment										
(I)	Power generation equipment										
1	Turbine equipment										
	Turbine (pump-turbine)										
	Governor										
	...										
2	Generator equipment										
	Generator (generator-motor)										
	Excitation device										
	...										
3	Inlet valve										
	Butterfly valve										
	...										
4	Hoisting equipment										
	Bridge crane										
	...										
5	Hydraulic machinery auxiliary equipment										

NB/T 10145-2019

Completed project – Table 03-2

Table B.12　Breakdown of transfer of usable fixed assets (equipment to be installed) *(continued)*

Unit: CNY

No.	Item description	Specification/ model	Supplier/ manufacturer	Installation location	Unit of measurement	Quantity	Equipment procurement cost	Installation works cost	Prepaid expenses	Asset value	Remarks
	1	2	3	4	5	6	7	8	9	10=7+8+9	11
(1)	Oil systems										
	Oil filter										
	…										
(2)	Compressed air system										
	Air compressors										
	…										
(3)	Water system										
	Water pump										
	…										
(4)	Hydraulic measurement system										
	Liquid level controller										
	…										
6	Electrical equipment										
(1)	Power generation voltage device										
	Generator circuit breaker										
	Current transformer										
	…										

NB/T 10145-2019

Completed project – Table 03-2

Table B.12　Breakdown of transfer of usable fixed assets (equipment to be installed) *(continued)*

Unit: CNY

No.	Item description	Specification/ model	Supplier/ manufacturer	Installation location	Unit of measurement	Quantity	Equipment procurement cost	Installation works cost	Prepaid expenses	Asset value	Remarks
	1	2	3	4	5	6	7	8	9	10=7+8+9	11
(2)	Variable frequency starting device										
	…										
(3)	Busbar										
	…										
(4)	Service power system										
	…										
(5)	Electrical test equipment										
	…										
(6)	Power cable										
	…										
(7)	Cable tray										
	…										
7	Control and protection equipment										
(1)	Control and protection system										
	…										
(2)	Computer supervision and control system										

NB/T 10145-2019

Table B.12 Breakdown of transfer of usable fixed assets (equipment to be installed) (continued)

Completed project – Table 03-2

Unit: CNY

No.	Item description	Specification/ model	Supplier/ manufacturer	Installation location	Unit of measurement	Quantity	Equipment procurement cost	Installation works cost	Prepaid expenses	Asset value	Remarks
	1	2	3	4	5	6	7	8	9	10=7+8+9	11
	…										
(3)	CCTV system										
	…										
(4)	DC system										
	…										
(5)	Control and protection cable										
	…										
8	Communication equipment										
(1)	Satellite communication										
	…										
(2)	Optical fiber communication										
	…										
(3)	Microwave communication										
	…										
(4)	Carrier communication										
	…										

NB/T 10145-2019

Completed project – Table 03-2

Table B.12 Breakdown of transfer of usable fixed assets (equipment to be installed) *(continued)*

Unit: CNY

No.	Item description	Specification/ model	Supplier/ manufacturer	Installation location	Unit of measurement	Quantity	Equipment procurement cost	Installation works cost	Prepaid expenses	Asset value	Remarks
	1	2	3	4	5	6	7	8	9	10=7+8+9	11
(5)	Mobile communication										
	...										
(6)	Production dispatch communication										
	...										
(II)	Step-up substation equipment										
1	Main transformer equipment										
(1)	Transformer										
(2)	Track										
(3)	Track stopper										
2	High voltage electrical equipment										
(1)	High voltage circuit breaker										
(2)	Current transformer										
(3)	Voltage transformer										
(4)	Disconnector										
(5)	Surge arrester										

Completed project – Table 03-2

Table B.12 Breakdown of transfer of usable fixed assets (equipment to be installed) *(continued)*

Unit: CNY

No.	Item description	Specification/ model	Supplier/ manufacturer	Installation location	Unit of measurement	Quantity	Equipment procurement cost	Installation works cost	Prepaid expenses	Asset value	Remarks
	1	2	3	4	5	6	7	8	9	10=7+8+9	11
(6)	GIS										
(7)	SF$_6$ gas GIL										
(8)	High voltage cable										
(9)	High voltage cable head										
3	Primary circuit line and others										
(III)	Dam-passing navigation equipment										
1	Power supply equipment										
	...										
2	Control equipment										
	...										
(IV)	Safety monitoring equipment										
	...										
(V)	Hydrologic, meteorological and sediment monitoring equipment										
	...										
(VI)	Fire protection equipment										

Table B.12 Breakdown of transfer of usable fixed assets (equipment to be installed) *(continued)*

Completed project – Table 03-2

Unit: CNY

No.	Item description	Specification/ model	Supplier/ manufacturer	Installation location	Unit of measurement	Quantity	Equipment procurement cost	Installation works cost	Prepaid expenses	Asset value	Remarks
1	2	3	4	5	6	7	8	9	10=7+8+9	11	
	...										
(Ⅶ)	Occupational health and safety equipment										
	...										
(Ⅷ)	Other equipment										
1	Lift equipment										
	Lift										
	...										
2	Feeder equipment in dam area										
	Transformer										
	...										
3	Water supply and drainage equipment										
	...										
4	Heat supply equipment										
	...										

Completed project – Table 03-2

Table B.12 Breakdown of transfer of usable fixed assets (equipment to be installed) *(continued)*

Unit: CNY

No.	Item description	Specification/ model	Supplier/ manufacturer	Installation location	Unit of measurement	Quantity	Equipment procurement cost	Installation works cost	Prepaid expenses	Asset value	Remarks
	1	2	3	4	5	6	7	8	9	10=7+8+9	11
5	Equipment apportioning cost of central control center of cascade power stations										
6	Ventilation and heating equipment										
(1)	Fan										
	Centrifugal fan (high pressure)										
	Glass fiber reinforced plastic centrifugal fan										
	...										
(2)	Air conditioner										
	Constant temperature and humidity										
	Screw type cooling unit										
	...										
(3)	Piping										
7	Earthquake monitoring station network equipment										

Completed project – Table 03-2

Table B.12 Breakdown of transfer of usable fixed assets (equipment to be installed) *(continued)*

Unit: CNY

No.	Item description	Specification/ model	Supplier/ manufacturer	Installation location	Unit of measurement	Quantity	Equipment procurement cost	Installation works cost	Prepaid expenses	Asset value	Remarks
	1	2	3	4	5	6	7	8	9	10=7+8+9	11
	...										
8	Plant-area earthing										
	...										
II	Hydraulic steel structures										
(I)	Water retaining works										
1	Gates										
	Radial gates										
	...										
2	Hoists equipment										
	Gantry crane										
	...										
3	Trash equipment										
	Trash rack										
	Trash rack embedded parts										
	...										
(II)	Discharge works										

Completed project – Table 03-2

Table B.12 Breakdown of transfer of usable fixed assets (equipment to be installed) *(continued)*

Unit: CNY

No.	Item description	Specification/ model	Supplier/ manufacturer	Installation location	Unit of measurement	Quantity	Equipment procurement cost	Installation works cost	Prepaid expenses	Asset value	Remarks
	1	2	3	4	5	6	7	8	9	10=7+8+9	11
1	Gates										
	...										
2	Hoists										
	...										
3	Trash racks										
	...										
(III)	Water conveyance works										
1	Gates										
	...										
2	Hoists										
	...										
3	Trash racks										
	...										
4	Steel penstocks										
	...										
(IV)	Powerhouse										

NB/T 10145-2019

Completed project – Table 03-2

Table B.12 Breakdown of transfer of usable fixed assets (equipment to be installed) *(continued)*

Unit: CNY

No.	Item description	Specification/ model	Supplier/ manufacturer	Installation location	Unit of measurement	Quantity	Equipment procurement cost	Installation works cost	Prepaid expenses	Asset value	Remarks
	1	2	3	4	5	6	7	8	9	10=7+8+9	11
1	Gates										
	…										
2	Hoists										
	…										
(V)	Step-up substation works										
	…										
(VI)	Navigation works										
1	Gates										
	…										
2	Hoists										
	…										
(VII)	Irrigation headworks										
	…										
	Total										
NOTE The breakdown items may be adjusted according to the relevant management regulations.											

Completed project – Table 03-3

Table B.13 Breakdown of transfer of usable fixed assets (equipment not to be installed)

Unit: CNY

No.	Items description	Specification/model	Supplier/ manufacturer	location or custody	Unit of measurement	Quantity	Asset value	Remarks
	1	2	3	4	5	6	7	10
I	Transport vehicles							
	...							
II	Mechanical repair equipment							
	...							
III	Test equipment							
	...							
IV	Tools							
	...							
V	Furniture							
	...							
VI	Observation and monitoring equipment							
	...							
VII	Others							
	...							
	Total							

59

Completed project – Table 03-4

Table B.14 Breakdown of transfer of usable assets (current assets)

Unit: CNY

No.	Items description	Specification/ model	Supplier/ manufacturer	location or custody	Unit of measurement	Quantity	Asset value	Remarks
	1	2	3	4	5	6	7	10
I	Current capital							
	…							
II	Stock materials handed over to production							
	…							
III	Stock equipment handed over to production							
	…							
IV	Cash and cash equivalents handed over to production							
	Bonds							
	Cash							
	…							
V	Spare parts							
	…							
	Total							

NB/T 10145-2019

Table B.15 Breakdown of transfer of usable assets (intangible assets and long-term prepaid expenses)

Completed project – Table 03-5

Unit: CNY

No.	Item description	Manufacturer	Location or user	Unit of measurement	Quantity	Asset value	Remarks
	1		2	3	4	6	8
I	Intangible assets						
	Total						
II	Long-term prepaid expenses						
	Total						

NOTE Documents and agreements related to the identification of intangible assets shall be indicated in the "remarks".

61

NB/T 10145-2019

Completed project – Table 04

Table B.16　Final financial account sheet

Unit: CNY

Sources of funds	Amount	Occupation of funds	Amount
Ⅰ. Project capital		Ⅰ. Total transfer of usable assets	
1.		Fixed assets	
2.		Current assets	
3.		Intangible assets	
4.		Long-term prepaid expenses	
Ⅱ. Infrastructure appropriation		Ⅱ. Infrastructure expenditures to be written-off	
1.		Ⅲ. Expenditures transferred out	
2.		Ⅳ. Input tax	
3.			
Ⅲ. Infrastructure loan		Ⅴ. Subtotal balance of funds	
1.		1. Equipment in stock	
2.		2. Materials in stock	
Ⅳ. Other funds		3. Currency funds	
1. Payables		4. Receivables	
2. Non-payments			
Total		Total	

62

Completed project – Table 04-1

Table B.17 Breakdown of yearly funds availability

Unit: CNY

No.	Item description	Year						Total
		1st year	2nd year	3rd year	4th year	5th year	…	
	Sources of funds							
1	Project capital							
	…							
2	Infrastructure appropriation							
	…							
3	Infrastructure loan							
	…							
4	Other funds							
	…							
	Total							

NB/T 10145-2019

Completed project – Table 05

Unit: CNY

Table B.18 Breakdown of wind-up works and reserve fund

| No. | Item description | Unit of measurement | Quantity | Project investment (input tax incl.) | Investment completed (input tax incl.) | Estimated uncompleted investment ||||| Remarks |
						Civil	Installation	Equip. Procurement	Others	Subtotal	Input tax	Total	
1	2	3	4	5	6	8	9	10	11	12=8+9+10+11	13	14=12+13	15
I	Project complex												
1	Construction auxiliary works												
…													
2	Construction works												
…													
3	Environmental protection works												
…													
4	Electro-mechanical equipment and installation works												
…													
5	Hydraulic steel structures and installation works												
…													
II	Resettlement compensation												
…													
III	Independent cost												
1													
…													
	Total												

Completed project – Table 06

Table B.19 Breakdown of prepaid expenses

No.	Item description	Prepaid expenses	Actual costs (excl. input tax)				Total
			Fixed assets	Current assets	Intangible assets	Long-term prepaid expenses	
	1	3	4	5	6	7	8=3+4+5+6+7
I	Construction auxiliary works						
1	Construction transport works						
2	Navigation works during construction						
3	Construction power supply works						
4	Construction water supply works						
5	Construction compressed air supply works						
6	Construction communication works						
7	Construction management information system						
8	Quarry/borrow area overburden stripping and protection works						
9	Aggregate processing system						
10	Concrete production and pouring system						
11	Diversion works						
12	Temporary safety monitoring works						
13	Temporary hydrologic telemetry and forecasting works						
14	Construction and management building works						
15	Other construction auxiliary works						
II	Construction works						

NB/T 10145-2019

Completed project – Table 06

Table B.19 Breakdown of prepaid expenses *(continued)*

| No. | Item description | Prepaid expenses | Actual costs (excl. input tax) | | | | Total |
			Fixed assets	Current assets	Intangible assets	Long-term prepaid expenses	
	1	3	4	5	6	7	8=3+4+5+6+7
1	Dam-passing navigation structures						
2	Abutment slope treatment works						
III	Environmental protection and soil and water conservation						
1	Environmental protection works						
2	Soil and water conservation works						
IV	Compensation for resettlement						
1	Impoundment-affected area						
2	Project complex construction area						
V	Independent cost						
1	Preliminary engineering fee						
2	Project construction management cost						
3	Resettlement management costs						
4	Project supervision costs						
5	Settlement monitoring and assessment costs						

NB/T 10145-2019

Completed project – Table 06

Table B.19 Breakdown of prepaid expenses *(continued)*

No.	Item description	Actual costs (excl. input tax)					Total
		Prepaid expenses	Fixed assets	Current assets	Intangible assets	Long-term prepaid expenses	
	1	3	4	5	6	7	8=3+4+5+6+7
6	Consulting service costs						
7	Technical and economic evaluation cost						
8	Cost for project quality inspection and test						
9	Management cost for norm standard preparation for hydropower project						
10	Acceptance cost						
11	Construction insurance fee						
12	Early mobilization cost for production staff						
13	Training cost						
14	Management tools procurement cost						
15	Spare parts procurement cost						
16	Cost for tools, instrument and production furniture procurement						
17	Joint commissioning fee						
18	Initial impoundment cost						

Completed project – Table 06

Table B.19　Breakdown of prepaid expenses *(continued)*

No.	Item description	Actual costs (excl. input tax)						Total
		Prepaid expenses	Fixed assets	Current assets	Intangible assets	Long-term prepaid expenses		
	1	3	4	5	6	7		8=3+4+5+6+7
19	Subsidy for grid-connected commissioning of units							
20	Research and test cost							
21	Investigation and design cost							
22	Other taxes and fees							
VI	Interest during construction period							
VII	Other costs							
1	Fixed assets loss							
2	Tools disposal loss							
3	Equipment inventory loss and damage							
4	Loss of scrapped works							
5	Damaged road repair cost							
6	Other prepaid expenses							
	Total							

Completed project – Table 07

Table B.20 Prepaid expenses apportionment calculation sheet

Unit: CNY

No.	Item description	Work quantities	Construction auxiliary works		Construction works			Environmental protection and soil and water conservation works		Compensation for land requisition and resettlement		Independent costs		Interest during construction period	Other costs		Total
			Construction transport works	...	Dam-passing navigation structure	...	...	Environmental Protection	Soil and water conservation	Impoundment-affected area	Complex area	Preliminary engineering fee	...		Fixed assets loss	...	
	1	2	3	4	5	6		7	8	9	10	11	12	13	14	15	16
	Prepaid expenses																
I	Structures																
1																	
2																	
3																	
4																	
II	Buildings																
1																	
2																	
III	Equipment to be installed																
1																	
2																	
3																	
4																	

Table B.21　Breakdown of infrastructure expenditures to be written-off

Completed project – Table 08　　　　　　　　　　　　　　　　　　　　　　　　　　　　　　　　　　Unit: CNY

No.	Category of expenditure	Amount	Reasons for write-off or approval document number
Total			

Table B.22　Breakdown of expenditures transferred-out

Completed project – Table 09　　　　　　　　　　　　　　　　　　　　　　　　　　　　　　　　　　Unit: CNY

No.	Item description	Location or characteristics	Unit	Quantity	Amount	Receiving entity
Total						

Annex C
(Informative)
Reference table for assets composition and prepaid assets classification for hydropower projects

Table C.1 specifies the actual assets composed of various investments and costs of a hydropower project as well as the major accounting entries of the prepaid expenses of the project.

Table C.1 Reference table for assets composition and prepaid assets classification

No.	Item description	Expenses directly identified as assets				Prepaid expenses	Remarks
		Fixed assets	Current assets	Intangible assets	Long-term prepaid expenses		
I	Project complex						
(I)	Construction auxiliary works						
1	Construction transport works	√				√	In the case of permanent and temporary use
2	Navigation works during construction	√				√	In the case of permanent and temporary use
3	Construction power supply works	√				√	
4	Construction water supply works	√				√	
5	Construction compressed air supply works					√	
6	Construction communication works					√	
7	Construction management information system					√	
8	Quarry/borrow area overburden stripping and protection works					√	
9	Aggregate processing system					√	

NB/T 10145-2019

Table C.1 Reference table for assets composition and prepaid assets classification *(continued)*

No.	Item description	Fixed assets	Current assets	Intangible assets	Long-term prepaid expenses	Prepaid expenses	Remarks
10	Concrete production and pouring system					√	
11	Diversion works					√	
12	Temporary safety monitoring works					√	
13	Temporary hydrologic telemetry and forecasting works					√	
14	Construction and management building works	√				√	Permanent and temporary combined
15	Other construction auxiliary works					√	
(II)	Construction works						
1	Water retaining (storage) works	√					
2	Discharge and energy dissipation structure	√					
3	Water conveyance structure	√					
4	Power generation structure	√					
5	Step-up substation structure	√					
6	Dam-passing navigation structure	√				√	River channel improvement costs belong to prepaid expenses
7	Irrigation canal head structure	√					
8	Abutment slope treatment works	√				√	Slope protection and clearance costs belong to prepaid expenses
9	Transport works	√		√			
10	Building works	√		√			

NB/T 10145-2019

Table C.1 Reference table for assets composition and prepaid assets classification (continued)

No.	Item description	Expenses directly identified as assets				Prepaid expenses	Remarks
		Fixed assets	Current assets	Intangible assets	Long-term prepaid expenses		
11	Safety monitoring works	√					
12	Hydrologic telemetry and forecasting works	√					
13	Fire protection works	√					
14	Occupational health and safety works	√					
15	Other works			√		√	
(III)	Environmental protection and soil and water conservation works						
1	Environmental protection works	√		√		√	
2	Soil and water conservation works	√		√		√	
(IV)	Electro-mechanical equipment and installation works	√					
(V)	Hydraulic steel structure equipment and installation works	√					
II	Compensation for resettlement						
1	Impoundment-affected area					√	
2	Project complex area			√		√	
III	Independent costs						
(I)	Project construction management costs						
1	Preliminary engineering fee	√		√		√	Fixed assets refer to equipment and tools procured by the owner
2	Owner's construction management cost	√				√	

73

Table C.1 Reference table for assets composition and prepaid assets classification *(continued)*

No.	Item description	Expenses directly identified as assets				Prepaid expenses	Remarks
		Fixed assets	Current assets	Intangible assets	Long-term prepaid expenses		
3	Resettlement management cost	√					
4	Construction supervision cost					√	
5	Resettlement monitoring and assessment cost					√	
6	Consulting service cost					√	
7	Technical and economic evaluation cost					√	
8	Cost for project quality inspection and test					√	
9	Management cost for norm standard preparation for hydropower project					√	
10	Project acceptance cost					√	
11	Construction insurance fee					√	
(II)	Operational production preparation fee						
1	Early mobilization costs for production staff				√		
2	Training cost				√		
3	Management tools procurement cost	√	√				
4	Spare parts procurement cost	√	√				
5	Cost for tools, instrument and production furniture procurement	√	√				
6	Joint commissioning fee					√	
7	Initial impoundment cost					√	
8	Subsidy for grid-connected commissioning of units					√	

Table C.1 Reference table for assets composition and prepaid assets classification *(continued)*

No.	Item description	Expenses directly identified as assets				Prepaid expenses	Remarks
		Fixed assets	Current assets	Intangible assets	Long-term prepaid expenses		
(Ⅲ)	Research, investigation and design cost						
1	Research and test cost					√	
2	Investigation and design cost					√	
(Ⅳ)	Other taxes and fees						
1	Arable land occupation tax					√	
2	Arable land reclamation cost					√	
3	Forest and vegetation restoration cost					√	
4	Soil and water conservation compensation fee					√	
5	Others					√	
Ⅳ	Interest during construction period					√	
Ⅴ	Others						
1	Fixed assets loss	√				√	
2	Equipment disposal loss	√				√	
3	Equipment inventory loss and damage					√	
4	Loss of scrapped works					√	
5	Damaged road repair cost					√	
6	Other prepaid expenses					√	

Annex D
(Informative)
Reference table for apportionment of prepaid expenses of hydropower projects

Table D.1 specifies the composition of the prepaid expenses for a hydropower project as well as the apportionment and apportionment methods.

Table D.1 Reference table for apportionment of prepaid expenses

No.	Item description	Accounting subjects	Subject apportioned to			Remarks
			Structure	Building	Equipment to be installed	
I	Project complex works					
(I)	Construction auxiliary works	Prepaid expenses	○	○	○	Excluding expenses constituting fixed assets
(II)	Civil works	Prepaid expenses	○			Excluding expenses constituting fixed assets
(III)	Environmental protection and soil and water conservation works	Prepaid expenses	○			Excluding expenses constituting fixed assets
II	Compensation for resettlement					
II-1	Impoundment-affected area	Prepaid expenses	○			
II-2	Project complex area	Prepaid expenses	○			Corresponding items
III	Independent costs					
(I)	Project construction management costs					
1	Preliminary engineering fee	Prepaid expenses	○	○	○	Excluding expenses constituting fixed assets
2	Owner's construction management cost	Prepaid expenses	○	○	○	Excluding expenses constituting fixed assets
3	Resettlement management cost	Prepaid expenses	○			Excluding expenses constituting fixed assets
4	Construction supervision cost	Prepaid expenses	○	○	○	

Table D.1 Reference table for apportionment of prepaid expenses *(continued)*

No.	Item description	Accounting subjects	Subject apportioned to			Remarks
			Structure	Building	Equipment to be installed	
5	Resettlement monitoring and assessment cost	Prepaid expenses	○			
6	Consulting service cost	Prepaid expenses	○	○	○	Excluding expenses of which the subject can be distinguished
7	Technical and economic evaluation cost	Prepaid expenses	○	○	○	
8	Cost for project quality inspection and test	Prepaid expenses	○	○	○	
9	Management cost for norm standardization preparation for hydropower project	Prepaid expenses	○	○	○	
10	Acceptance cost	Prepaid expenses	○	○	○	
11	Construction insurance fee	Prepaid expenses	○	○	○	By the subject of insurance
(II)	Operational production preparation fee					
1	Joint commissioning fee	Prepaid expenses			○	
2	Initial impoundment cost	Prepaid expenses	○	○	○	
3	Subsidy for grid-connected commissioning of units	Prepaid expenses			○	
(III)	Research, investigation and design cost					
1	Research and testing cost	Prepaid expenses	○	○	○	Excluding expenses of which the subject can be distinguished
2	Investigation and design cost	Prepaid expenses	○	○	○	
(IV)	Other taxes and fees					
1	Arable land occupation tax		○			
2	Arable land reclamation cost		○			

Table D.1 Reference table for apportionment of prepaid expenses *(continued)*

No.	Item description	Accounting subjects	Subject apportioned to			Remarks
			Structure	Building	Equipment to be installed	
3	Forest and vegetation restoration cost		○			
4	Soil and water conservation compensation fee		○			
5	Others		○			
IV	Interest during construction period	Prepaid expenses	○	○	○	
V	Others					
1	Fixed assets loss	Prepaid expenses	○	○	○	
2	Equipment disposal loss	Prepaid expenses	○	○	○	
3	Equipment inventory loss and damage	Prepaid expenses	○	○	○	
4	Loss of scrapped works	Prepaid expenses	○	○	○	
5	Damaged road repair cost					
6	Other prepaid expenses	Prepaid expenses	○	○	○	